GETTING ENOUGH SLEEP

By R. J. MacReady

New York

Published in 2022 by Cavendish Square Publishing, LLC
243 5th Avenue, Suite 136, New York, NY 10016

First Edition

Library of Congress Cataloging-in-Publication Data

Names: MacReady, R. J., author.
Title: Getting enough sleep / R.J. MacReady.
Description: First. | New York : Cavendish Square Publishing, [2022] |
Series: Healthy choices | Includes index.
Identifiers: LCCN 2020034845 | ISBN 9781502659606 (library binding) | ISBN 9781502659583 (paperback) | ISBN 9781502659590 (set) | ISBN 9781502659613 (ebook)
Subjects: LCSH: Sleep–Juvenile literature.
Classification: LCC QP425 .M23 2022 | DDC 612.8/21–dc23
LC record available at https://lccn.loc.gov/2020034845

Editor: Greg Roza
Designer: Andrea Davison-Bartolotta

The photographs in this book are used by permission and through the courtesy of: Cover Hung Chung Chih/Shutterstock.com; p. 5 (top left) Paul Zinken/picture alliance via Getty Images; p. 5 (top right) www.anitapeeples.com/Moment/Getty Images; p. 5 (bottom left) Andrea Ricordi, Italy/Moment/Getty Images; p. 5 (bottom right) Tetra Images/DigitalVision/Getty Images; p. 7 Tim Pannell/Corbis/VCG/Corbis/Getty Images Plus/Getty Images; p. 9 Mark Edward Atkinson/Tracey Lee/Getty Images; p. 11 Klaus Vedfelt/DigitalVision/Getty Images; p. 13 OJO Images/Getty Images; p. 15 Oksana Kuzmina/Shutterstock.com; p. 17 Thanasis Zovoilis/Moment/Getty Images; p. 19 romrodinka/iStock/Getty Images Plus/Getty Images; p. 21 Jose Luis Pelaez Inc/DigitalVision/Getty Images; p. 23 rogkov/E+/Getty Images.

CPSIA compliance information: Batch #CW22CSQ: For further information contact Cavendish Square Publishing LLC, New York, New York, at 1-877-980-4450.

Printed in the United States of America

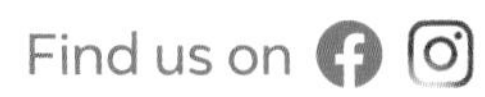

CONTENTS

Everyone Needs to Sleep

Most animals on Earth sleep in some way. Some animals sleep up to 20 hours a day! People don't need that much sleep, but sleep is very important to our health.

Healthy people are active, which means they're busy! Kids go to school. They play and do **chores**. Adults go to work. We use **energy** doing these things. We eat food to get energy. We also need sleep after a busy day.

Sleep lets our bodies **recharge**! That means that when we sleep, our **muscles** get to rest and grow. Sleep helps injuries, such as cuts, heal faster. Sleep gets your body ready for a new day.

Sleep lets our brains, or minds, recharge too! Our brains are busy during the day. Sleep gives the brain a chance to rest. Sleep helps us remember what we learned during the day.

People who don't get enough sleep will feel tired during the day. They may have a hard time thinking. People who don't sleep enough get sick more often. Getting enough sleep helps our bodies stay strong to fight illnesses.

Time to Sleep

How much sleep do you need? It depends on your age. Babies need more than 12 hours of sleep. School-age kids need about 10 hours. Teens need nine hours. Adults need about eight hours of sleep.

When you first fall asleep, you're still easy to wake up. After some time, your brain tells your muscles to rest. It tells your heart to slow down. You breathe slower. Soon, you're in a deep sleep.

The brain gets active after deep sleep. That's when we dream! Everyone dreams. We don't always remember them. Dreams often say something about how we feel. If you're sad about something, you might have a sad dream.

Your eyes move a lot when you dream. This is called **rapid** eye movement, or R.E.M. Your heart may beat faster too. R.E.M. sleep doesn't last all night. You go between the different kinds of sleep until it's time to wake up.

Healthy Sleep Choices

You can make healthy sleep choices. You can exercise every day. This tires you out! It's a good idea to turn off your TV and put down your phone. They keep you awake. Many people like to read before bed.

WORDS TO KNOW

chores: Jobs at home.

energy: The ability to be active.

muscles: Body parts that allow a person to move.

rapid: Very fast.

recharge: To regain energy.

INDEX